Tri-Color
The De Facto Standard for Dynamic Containment Testing

by

Chip Albright

FUME HOOD CERTIFIED

A Fume Hood Certified Publication

Fume Hood Certified, LLC
PO Box 71477
Phoenix, AZ 85850 USA
www.fumehoodcertified.com
experts@fumehoodcertified.com
+1 623-662-0907

Disclaimer

This publication is designed to provide accurate and authoritative information with regard to the subject matter covered. It is sold with the understanding that the contents do not render professional advice. If you require professional assistance or expert advice, you should seek the services of a licensed professional.

Any company names, logos, or images used that are trademarked or copyrighted by others are acknowledged to be the property of the trademark or copyright owner.

This book was created with assistance from artificial intelligence resources. The content has been reviewed and edited by a human. For more information on the extent and nature of AI usage please contact Chip Albright.

Paperback: ISBN 978-1-7357110-3-4
First paperback edition December 2025.

Printed in the USA

Table of Contents

About the Author

When Chip Albright walked into his first fume hood factory in the mid-1980s, he had no idea he was stepping into what would become a lifelong mission. At Kewaunee Scientific in Lockhart, Texas, Chip was handed a clean sheet of paper and asked to help design better hoods. He threw himself into the work, quickly realizing these weren't just pieces of steel and glass—they were life-saving devices, the first line of defense for countless scientists.

Chip moved through the ranks of industry leaders like Jamestown-JMP and Collegedale LLC, always chasing the same question: How can we make fume hoods safer? He designed seven major product lines and watched hundreds of thousands of his hoods get installed in laboratories worldwide. Each project taught him something new about the relationship between hood design, laboratory ventilation systems, and how people actually work at the bench.

In 1999–2000, as Chairman of SEFA, he led the rewrite of SEFA-1: Fume Hoods, a turning point in setting global standards. He became known not just as an engineer, but as a voice of reason in an industry full of myths, challenging the long-held belief that face velocity was the ultimate measure of safety. To Chip, it was always about containment.

That message resonated far beyond the United States. In China and India, Chip became a highly sought-after speaker and advisor. His Tri-Color Laser Testing workshops in India drew crowds of EHS professionals and engineers eager to see how airflow really behaved inside a hood. In China, he worked alongside manufacturer Sunway to reimagine what a "smart hood" could be, helping shape new generations of technology less dependent on perfect room conditions and more focused on user protection.

His book Laboratory Fume Hoods Explained and his workshops, white papers, and presentations carry the same voice—practical, unvarnished, and deeply informed by years on factory floors, in labs, and at test benches. Through Fume Hood Certified LLC, Chip has trained thousands, from lab managers to architects to young engineers, always reminding them that the hood is not just equipment—it's a lifeline. Today, Chip's mission remains unchanged: "Making labs safer, one fume hood at a time."

Introduction

How We Got Here

Every day, countless scientists work at fume hoods believing a simple idea: if the face velocity reads 100 feet per minute, they're safe.

They're wrong.

Air Speed ≠ Containment

This misconception has persisted for decades, protected by familiarity and reiterated by many industry professionals. Containment failures happen during routine work when someone walks by, when a door opens, when the sash moves. These dynamic events aren't captured by traditional testing, yet they're when exposure most often occurs.

The current gold standard, ASHRAE 110, provides valuable baseline data but tests hoods under static, controlled conditions with the exception of the sash movement containment test. It even acknowledges this limitation, suggesting that "performance under dynamic conditions" should be evaluated.

Meanwhile, laboratories continue to rely on face velocity measurements—a metric that tells us how fast air moves but nothing about whether contaminants are actually contained. It's like measuring a car's speed to determine if it's safe to drive, ignoring whether the brakes work or the airbags deploy.

In the tester world there is a name for a tester who goes in with a handheld annometer takes a face velocity reading and releases a little smoke and then slaps on a safety sticker saying the hood is performing safely.

Time to change the discussion from face velocity to containment

The Tri-Color Solution

Tri-Color laser visualization changes this equation. Using three colors of laser light and a harmless haze, it makes containment visible in real time:

- Blue reveals airflow patterns inside the hood
- Red exposes dangerous turbulence at the inside of the sash plane
- Green creates a containment barrier that instantly shows when potentially hazardous materials escape

No interpretation needed. No complex calculations. If haze crosses the green line, containment has failed. Anyone can see it, understand it, and act on it.

Why This Book Exists

After years of development and thousands of hood tests worldwide, Tri-Color has proven itself as a practical, accessible method for validating dynamic containment. Yet it remains absent from most specifications, unknown to many safety professionals, and unutilized in facilities that need it most.

This book makes the case that Tri-Color is the de facto standard for dynamic containment testing. Not to replace existing standards, but to fulfill their original intent: ensuring that fume hoods actually protect the people who depend on them.

The chapters ahead trace this journey from the physics of airflow to the future of smart hoods. But the message is simple: when safety is visible, safety is undeniable. And in today's laboratories, anything less than visible proof of containment is no longer acceptable.

TESTED USING
TRI-COLOR
FUME HOOD VISUALIZER

Chapter 1
The History of Laser Visualization

The First Light in the Dark
The invisible world of airflow inside a fume hood has always been difficult to understand. For decades, testers relied on smoke—messy, white-on-white, and often misleading. In 1991, Dr. Rong Fung Huang at the National Taiwan University of Science and Technology had a radical idea while earning his PhD in the US: use lasers to see the air itself.
By cutting haze with a sharp beam of light, he revealed vortices, turbulence, and flows never visible before. His insights led to patented designs using air curtains to reshape airflow, converting chaotic three-dimensional movement into more stable two-dimensional patterns. Without lasers, those designs would never have been possible.

Milestone #1
1991: Dr. Huang pioneers laser airflow visualization in fume hoods. Result: Patented designs with air curtains, reduced turbulence, improved containment.

A Spark in the West
I first encountered Dr. Huang's work in 2007. At the time, I was working on my third-generation fume hood design. His published article on sash movement and walk-by leakage opened my eyes. I had never seen airflow like that before.

The first time I tried laser visualization myself, I realized much of what I thought I knew was wrong. Lasers revealed turbulence and pressurization in ways that instruments could never quantify. Problems I couldn't explain with velocity numbers suddenly made sense when I could see the air moving.

There was a catch, for lasers to move beyond R&D into everyday testing, the process needed to be simplified, standardized, and repeatable.

Insight: Seeing the airflow for the first time was like looking through a microscope. Once you see the hidden world, you can't unsee it.

From Smoke to Haze

Traditional smoke wasn't enough. Fog machines were messy, hot, and disruptive. They injected turbulence of their own, masking the patterns we needed to observe. Haze machines (cooler, gentler, less dense) proved far superior. Unlike fog, haze flowed with the air instead of fighting it.
When the haze met the laser, the airflow came alive. The invisible was visible.

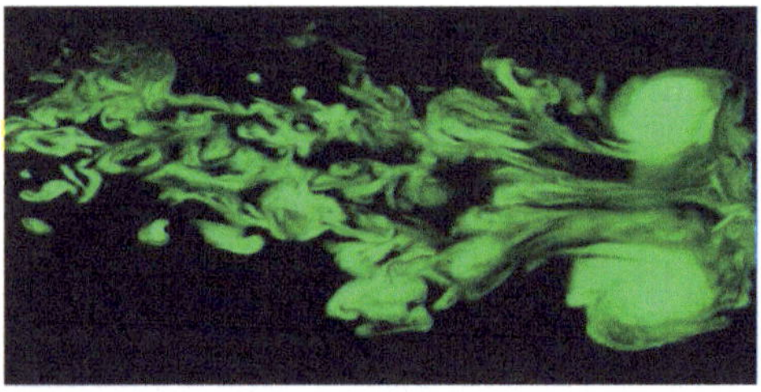

The Tri-Color Breakthrough

By 2007, I began experimenting with combining haze and multiple laser colors. The goal was to create a system anyone could understand at a glance. After years of testing, the Tri-Color protocol emerged. This was the moment airflow visualization shifted from a specialist's tool to a democratized safety test.

Milestone #2

2007: Early Tri-Color experiments begin.

Result: Visual containment testing that any user can understand.

The Takeaway

Laser visualization isn't new—it's been shaping hood design for more than 30 years. With Tri-Color, lasers moved from the realm of innovation and patents into everyday testing, education, and safety assurance. You can't fix what you can't see. Lasers gave us vision. Tri-Color gave us clarity.

Chapter 2
ASHRAE 110 and Dynamic Containment

A Standard That Changed Everything
Before 1985, fume hood testing was the Wild West. Each manufacturer claimed their hood was the best, but there was no common yardstick. Buyers compared face velocity numbers in marketing brochures, but neither told the real story of containment.

ASHRAE 110 in 1985 changed that. For the first time, laboratories had a standardized method to evaluate fume hood performance. It introduced:

- **Face velocity profile testing** - to examine turbulence across the sash opening
- **Smoke visualization** - to observe containment patterns
- **Tracer gas testing with SF_6** - to quantify leakage using mannequins and detectors

It wasn't a performance standard—there was no pass/fail. It was a test method that gave laboratories a way to measure containment rather than assume it.

> **Key Insight:** ASHRAE 110 is not a safety guarantee. It's a test method designed to provide data. Safe performance depends on interpreting that data in context.

AM, AI, and AU - Three Views of Containment

ASHRAE 110 created three distinct tests:

- **AM (As Manufactured)** - performed in a perfect test room. Reveals the hood's design potential, isolated from room conditions.
- **AI (As Installed)** - performed in the actual lab, with building systems operating. Shows how the hood interacts with its environment.
- **AU (As Used)** - performed under real working conditions, with equipment, chemicals, and user behavior included.

These tests remind us that a hood doesn't operate in a vacuum. Containment is always influenced by the room, the system, and the user.
The AU Test: The closest thing we have to real life. It's messy, dynamic, and often the most revealing.

The Standard's Quiet Suggestion

Buried in ASHRAE 110 is a line that would later change the industry:

"It is important to evaluate the performance of the laboratory hood under dynamic conditions. This performance test method may be modified to evaluate a dynamic challenge."

This wasn't a mandate, but it was a door left open. ASHRAE acknowledged that hoods live in a dynamic world. People walk by. Doors open. Sashes move. Cross drafts shift. All of these disrupt containment, yet the standard's prescribed tests only scratched the surface.

For years, most ignored this invitation. Dynamic testing was considered too complicated, too expensive, too difficult to standardize. For some of us, it was the most important part of the entire document.

When Standards Lag, Innovation Leads

Standards take years to update. Innovation moves faster.

- ASHRAE 110 has evolved slowly
- Tri-Color testing already addresses challenges not covered in current standards
- Waiting for standards means delaying safety improvements

Tri-Color adoption has identified containment risks missed by legacy SF6 testing. While not a standard written by an association, the Tri-Color testing protocol has been accepted as an enhancement to ASHRAE 110

These traffic lights are not standard, but they are innovative and are proving to be safer and more effective at preventing accidents. Someday after enough people have adopted them, they will likely become a standard but for now they are just a de facto standard.

Tri-Color adoption has identified containment risks missed by legacy SF_6 testing. While not a standard written by an association, the Tri-Color testing protocol has been accepted as an enhancement to ASHRAE 110.

Dynamic Challenges:

- Full Open Sash
- Design Open of the sash
- Rapid movement within the hood
- Objects being removed from the hood
- Pedestrian walk-bys
- Door openings
- Rapid sash movement
- Cross drafts

- Other hoods in the room being opened and closed

These aren't exceptions. They're the daily reality of lab life.

SF_6 and Its Limits

The tracer gas test with sulfur hexafluoride (SF_6) became the gold standard for quantifying containment. The equipment is highly sensitive, capable of detecting leakage down to parts per billion. SF_6 came with problems:

- Potent greenhouse gas
- Requires specialized (and expensive) detection equipment
- Measures a snapshot in time, not ongoing performance
- Its behavior inside the hood isn't always representative of real-world contaminants

We don't know what we don't know. Even with thousands of tests, questions remain about how SF_6 challenges the hood and whether it tells the full story of containment.

The Gap That Needed Filling

ASHRAE 110 was a milestone, but it left laboratories with a blind spot. The tests were controlled, static, and often disconnected from the chaos of daily lab work. The dynamic challenge clause was there, waiting for someone to take it seriously.

Tri-Color enters the story here.
Built on the foundation of ASHRAE 110, Tri-Color does exactly what the standard suggested: it evaluates performance under dynamic conditions. It doesn't replace ASHRAE 110—it fulfills its intent.

By combining haze, lasers, and simple visual cues, Tri-Color turns dynamic containment testing into something accessible, repeatable, and undeniable. Anyone watching can see when a hood holds or when it fails.

Closing Thought

ASHRAE 110 gave us the foundation. SF_6 gave us precision. The real world gave us turbulence, cross drafts, and chaos. The next chapters show how Tri-Color builds on that foundation—transforming a quiet suggestion in the standard into a new de facto practice for dynamic containment testing.

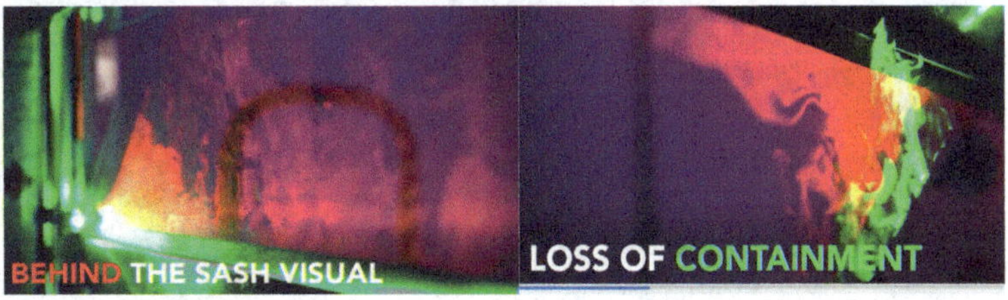

Chapter 3
How Tri-Color Works

Making the Invisible Visible

For years, the challenge of testing fume hoods has been invisibility. Air is clear. Contaminants are often invisible. Even with smoke or tracer gas, interpretation requires specialized equipment or expert eyes.

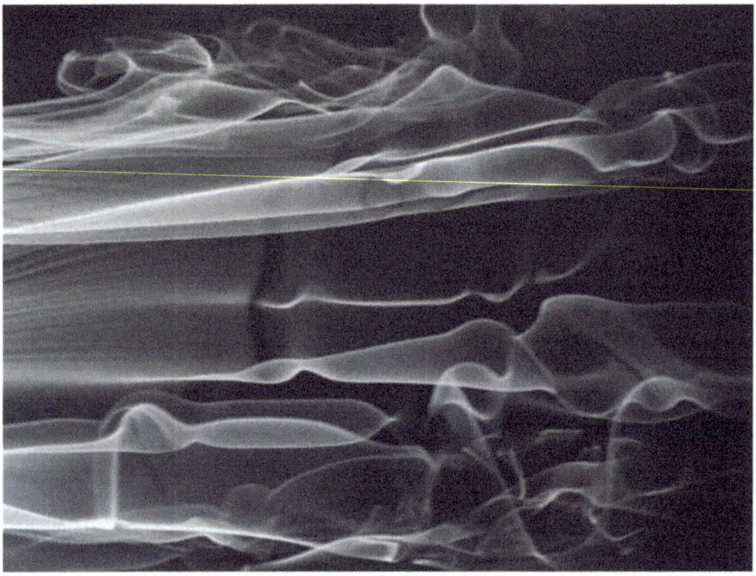

Tri-Color was designed to solve that problem. It takes the complex science of airflow and translates it into a visual language anyone can understand: blue, red, and green light.

The Tri-Color Promise:

- Simple enough for routine testing
- Visual enough for anyone to see
- Repeatable and recordable with video
- Built on ASHRAE 110's dynamic challenge principle

The Setup

At the heart of Tri-Color is a combination of:

- **Lasers** – fixed beams that cut through the airflow
- **Haze** – a lightweight, low-velocity mist that reveals movement without disrupting it
- **Video capture** – documentation that preserves what the human eye sees

The haze machine is critical. Unlike fog, which is hot and turbulent, haze is cool and gentle. It flows naturally with the hood's airflow instead of fighting it. What you see is a true representation of containment, not an artifact of the test.

The Three Colors

Each color tells a different part of the story:

Blue Beam – The Cross-Section of Flow

The blue laser slices vertically through the interior of the hood chamber. It shows the general airflow pattern and whether air is sweeping evenly across the space or forming dead zones. As a cross-section, it best shows the vortex.

Red Beams - Turbulence Inside the Sash
The red lasers illuminate instability where user and hood interact. Swirls, vortices, and rolling eddies glow red, making turbulence impossible to ignore. Loss of containment often occurs when the vortex collapses.

Green Curtain - The Containment Line
The most powerful feature. The green lasers form a glowing curtain at the sash opening. This is the hood's defensive wall. If haze crosses that line, containment has failed. Why green on the outside when red is inside? Green lasers biologically provide the most contrast to the eye, making it easier to see micro failures.

Why It Works

Tri-Color takes what ASHRAE 110 only suggested—dynamic containment testing—and makes it practical. Instead of a single detector number, you see the entire event. Instead of relying on interpretation by experts, you watch a story unfold in color and motion that anyone can understand.
It's not just a test. It's an educational tool. Lab workers who see their own actions cause turbulence or breaches never forget the lesson. The cause-and-effect clarity is unmatched.

Closing Thought

With Tri-Color, the hood speaks for itself. The blue beam shows how air flows. The red beams expose instability. The green curtain declares whether containment holds. No graphs, no equations, no debate. Just light, haze, and truth.

This is why Tri-Color is rapidly becoming the de facto standard for dynamic containment testing.

Chapter 4
Visualization: Changing Our Understanding

When Seeing Beats Measuring

For decades, laboratories relied on numbers to define safety. Average face velocity, tracer gas concentrations, flow rates. Numbers feel precise. They give the illusion of certainty.

If there's an accident and legal action ends up in court, which would you rather have? A test report with numbers or a video of actual performance? Numbers can fail to tell the whole story. A hood may have a perfect 100 fpm average face velocity and still leak contaminants into the room. A tracer gas test might produce a passing report while turbulence swirls directly through a user's breathing zone.

Visualization changes that. When you can see containment loss—haze glowing green as it crosses the sash curtain—there's no debate, no footnote, no "acceptable variation." It's happening, right in front of you.

You still need a risk-based assessment of the danger of exposure. With some chemicals, a micro failure might be acceptable; with others, not. That's no different than saying 0.10 PPM is acceptable but 0.11 PPM is a failure. Nothing about this is black and white.

The Face Velocity Myth:
Face velocity measures speed of air
It does not measure containment

The Aha! Moment

When we first beta-tested Tri-Color at universities, the reactions were immediate:

"I've been working with hoods for years and I've never seen anything like this...This should be part of every training program."

Why such a strong response? Because visualization provides what numbers cannot: context.

- You see how a careless arm motion pulls haze out of the hood
- You see how a nearby door draft bends the green containment curtain
- You see turbulence glowing red just where a flask is placed too close to the sash

These insights aren't theoretical. They're real, repeatable, and undeniable.

With a well-performing hood, you'll likely see some loss of containment (green). You might see micro failures or even massive failures. Tri-Color is your first line of defense. Depending on what you're doing and the amount and frequency of the green, you may need to do additional troubleshooting to determine how bad the problem is and what can be done to fix it.

Seeing is Believing: Visualization doesn't just test containment. It teaches it.

Numbers Without Vision

Consider the ASHRAE 110 tracer gas test. Highly sensitive. Quantitative. The report might say "0.05 ppm." What does that mean to the average user? Most can't interpret it. Even many testers don't know how to translate a result into corrective action.

It's like an X-ray. A technician can take the picture, but only a radiologist can explain what it means. Visualization flips this model. With Tri-Color, the meaning is clear to anyone who can see.

From Data to Dialogue

Visualization changes the conversation.

> Numbers spark debate:
> "Is 0.10 ppm good enough?"
> "Does 100 fpm guarantee safety?"
> "What's an acceptable face velocity range?"

Visualization sparks action:

> "That haze should not be crossing the green curtain."
> "That turbulence is a problem we can fix."
> "That door draft is causing loss of containment."

Visualization replaces argument with agreement.

Why This Matters

Fume hoods are exposure control devices. Their entire purpose is to keep hazards inside the fume chamber. Anything that makes containment failure obvious helps us fulfill that mission.

Tri-Color isn't just a test—it's a way of changing minds. It shifts safety from compliance ("did we pass the test?") to prevention ("is the user protected?"). It trains students, reassures researchers, and guides facilities managers toward better design and operation.

Closing Thought

When you see it, you understand it. Visualization doesn't replace standards like ASHRAE 110—it fulfills them. It gives clarity where numbers give confusion. And in that clarity lies the power to make labs safer, one hood at a time.

Fume Hood Performance Testing is not just a checkbox for compliance. It's validating that the fume hood is doing what it was designed to do: containing chemicals and preventing user exposure.

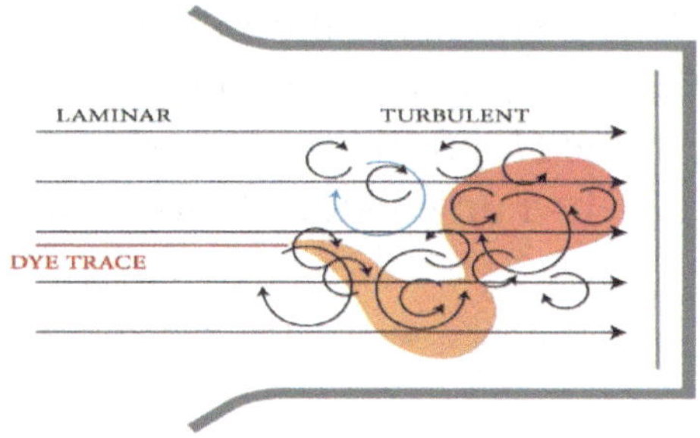

Chapter 5
Building on ASHRAE 110

A Solid Foundation

ASHRAE 110 has been the backbone of fume hood testing since 1985. It gave laboratories a common language to measure performance—face velocity profiles, smoke visualization, tracer gas quantification. But as the 2016 edition makes clear, many variables affect a hood's ability to protect: cross drafts, sash movement, thermal loads, and user practices.

Tri-Color builds on this foundation. What began as a way to make airflow visible has matured into a structured, repeatable protocol—one that complements ASHRAE 110 while delivering the dynamic clarity it only suggested.

The Tri-Color Protocol: Step by Step

Step 1: Preparation & Conditions

Like ASHRAE 110, environment matters. Confirm HVAC systems are in normal operation. Note cross drafts and diffuser types. Position the hood "as used," not just pristine. Document with photos—these become part of the Hood Identification Number (HIN) record in the Fume Hood Performance Tracker.

Video Capture and Documentation
This is where Tri-Color becomes undeniable. A 20-page velocity report might be ignored; a video of haze crossing the green curtain ends the debate instantly. Best practices:

- Always use a tripod; keep the sash and curtain framed.
- Dim background light to sharpen laser contrast.
- Capture at least 30 seconds per test.
- Narrate or label clips ("walk-by test," "rapid sash movement")

Upload all media to the hood's HIN record in the Fume Hood Performance Tracker. Reports can include screenshots, QR codes linking to full videos, or embedded clips—transforming static documents into living records.

Video doesn't just prove compliance; it trains. Students learn instantly when a hand wave causes a breach. EHS officers build a library of "what to do" and "what not to do." Managers gain leverage when requesting repairs or upgrades. In court, video speaks louder than numbers ever could.

Step 2: Face Velocity Profile

Divide the sash into a grid, place a hot-wire anemometer probe in each cell, and record 20 readings. Average results expose turbulence patterns. Remember: velocity is speed, not containment. A stable profile suggests consistency, but only visualization shows whether hazards stay inside.

Step 3: Airflow Visualization

Using a zero-velocity haze generator, challenge the hood at multiple points.

- *Small volume tests*: airfoil, sash edges, sidewalls.
- *Large volume tests*: chamber interior, work surface, sash plane.

  Observe: does haze sweep smoothly in, or does it roll back and linger? Document with photos and video.

Step 4: The Tri-Color Laser Setup

Introduce the three beams:

- *Blue beam*: airflow across the chamber interior.
- *Red beams*: turbulence in the sash plane.
- *Green curtain*: the containment barrier.

The green curtain is critical—if haze crosses it, containment has failed. Position cameras off to the side of the sash for the clearest view.

Step 5: Dynamic Challenges

ASHRAE 110 quietly recommended it; Tri-Color makes it real. Simulate real-world use:

- Open/close the sash at different speeds
- Pedestrian walk-bys
- Door openings
- Rapid hand movement or removing objects
- Neighboring hoods opening/closing

Each action reveals how fragile—or resilient—containment really is.

Complement, Not Replace

Tri-Color doesn't compete with ASHRAE 110. It completes it.

ASHRAE 110 delivers baseline numbers.
Tri-Color delivers visible proof.

Together, they create a holistic record of performance: rigorous data plus undeniable visualization.

Why This Matters

The best test is the one that drives action. Numbers sit in binders; videos drive change. Architects, EHS professionals, and users grasp Tri-Color instantly—because they can see it.
ASHRAE 110 laid the foundation. Tri-Color builds the house, filling the gaps on dynamic challenges, turning invisible risks into visible lessons, and preserving proof that no one can ignore.

This is containment testing anyone can understand, and everyone can trust.

Chapter 6
The Tri-Color Protocol

The Benchmark: SF_6

Since 1985, sulfur hexafluoride (SF_6) has been the tracer gas of choice in ASHRAE 110 testing. It's stable, detectable in parts per billion, and highly effective for quantifying containment. For decades, SF_6 data has been the benchmark against which fume hoods are judged.

SF_6 has serious limitations:

- **Environmental impact** – one of the most potent greenhouse gases known
- **Cost and availability** – expensive, sometimes restricted, difficult to obtain in many regions
- **Complexity** – requires calibrated detectors, ejectors, and skilled technicians
- **Snapshot results** – produces numbers, but no visual evidence, and no context for dynamic conditions

SF_6 Strengths & Weaknesses:

✓ High sensitivity
✓ Quantitative precision
✗ Expensive
✗ Greenhouse gas
✗ Complex test setup
✗ Results often misunderstood

The Tri-Color Alternative

Tri-Color was designed to offer a practical complement—and for many applications, an alternative—to SF_6 testing. Instead of tracer gas, it uses cool haze and laser visualization to provide immediate, visible proof of containment performance.

- Haze reveals airflow patterns without disrupting them
- Blue, red, and green lasers show stability, turbulence, and containment breach
- Video capture records events in real time

The difference is simple: SF_6 tells you a number. Tri-Color shows you what happened.

The Equivalency Question

For Tri-Color to be taken seriously, it had to prove equivalency. Side-by-side studies were conducted: the same hood tested with SF_6 and with Tri-Color. The results were clear:

- Hoods that leaked under SF_6 also showed haze escaping past the green curtain. The larger the leak, the more green was present.
- With Tri-Color you could see where the leakage was occurring and for how long.
- Hoods that contained under SF_6 showed no leakage with Tri-Color.

An unexpected result: with Tri-Color we saw leakage through the bypass that was pulled down through the user's breathing zone and back into the hood. Yet SF_6 indicated no leakage. Further testing showed that since SF_6 is 6 times heavier than air, it tends to fall to the work surface, leaving the upper hood relatively unchallenged.

Haze, by contrast, is entrained into the airflow including the vortex, showing leaks and recirculation zones that gas detectors cannot see.

Seeing vs. Measuring:

- SF_6 measures concentration at a single point
- Tri-Color visualizes the entire sash opening

Practical Advantages

For routine hood testing, ASHRAE 110 doesn't recommend using SF_6. On As-Used tests, you're not actually testing containment.

Compared to SF_6, Tri-Color offers advantages that SF_6 cannot:

- **No greenhouse gas release** - environmentally responsible
- **Simple and affordable** - haze and lasers are far less costly than gas cylinders and detectors
- **Accessible to more testers** - doesn't require years of experience and training

- **Dynamic testing** - shows containment during real-world challenges like sash movement and walk-bys

SF_6 remains valuable for R&D and precision benchmarking, but for day-to-day assurance, Tri-Color provides clarity and confidence.

Lessons from History

We don't know what we don't know. For decades, laboratories trusted SF_6 results without fully understanding how the gas behaved inside the hood. Seeing SF_6 with an infrared camera proved just how unpredictable its plume could be.
Tri-Color eliminates that blind spot. When containment fails, you don't need a number—you see it, instantly.

Closing Thought

SF_6 gave us a benchmark. Benchmarks aren't enough. Labs need methods that are accessible, repeatable, and trusted by those who depend on them. Seeing is believing.

Tri-Color doesn't replace SF_6—it builds on it, offering a greener, simpler, and more dynamic way to prove containment. With thousands of hoods tested with Tri-Color, it's rapidly being recognized as the de facto standard for dynamic containment testing

Chapter 7
The Future of Tri-Color

From Tool to Standard

When Tri-Color began, it was an experiment: a handful of lasers, a haze generator, and the question, "Can we make airflow visible in a way anyone can understand?"—simple and inexpensive.

Today, it has become a protocol. Tomorrow, it will be a standard. Beyond that, Tri-Color has the potential to become part of the everyday DNA of laboratory design, testing, and operation.

The future isn't just about better testing—it's about building a safer laboratory ecosystem where containment is continuously visible, verifiable, and trusted.

The Trajectory:

→ Experiment
→ Protocol
→. Specification
→ Standard
→ Ecosystem

Integration with Smart Hoods

Manufacturers are already embedding sensors, monitors, and digital dashboards into their "smart" hoods. Tri-Color is a natural fit:

Self-validation modules - haze and lasers built directly into the hood, activated with the push of a button

Digital logging - test results uploaded automatically to a secure cloud record

User feedback - sash light indicators that warn when turbulence or breaches occur

Imagine a hood that can test itself in real time and train users by showing them exactly how their actions affect containment.

AI and Automated Analysis

The next step is intelligence. With advances in AI-based video recognition, the future of Tri-Color includes:

- Automated detection of breaches in the green curtain
- Scoring of turbulence intensity in the red beams
- Pattern recognition to flag recurring containment issues

Instead of relying solely on human eyes, AI can interpret the videos, compare them against historical data, and generate actionable insights.

Toward Global Standards

As ASHRAE 110 and EN 14175 evolve, committees are wrestling with the challenge of dynamic testing and alternatives to SF_6. Tri-Color is well positioned to fill that gap.

North America – integration into modified ASHRAE 110 reports

Europe – visual complement to EN 14175's containment philosophy

Asia – rapid adoption in India and China, where cost and accessibility drive innovation

A future global standard could include both quantitative tracer gas testing and qualitative Tri-Color visualization, working together to deliver a complete picture of containment.

Beyond the Hood

Tri-Color's future isn't limited to chemical fume hoods. The same principles apply to:

- Biosafety cabinets
- Local exhaust systems
- Cleanrooms
- Isolation rooms

Wherever airflow matters, visualization can bring clarity.

Closing Thought

Tri-Color started as a way to make the invisible visible. Its future is about making safety continuous, intelligent, and global.

The day will come when pushing a "magic button" on the hood triggers a Tri-Color self-test, records the results, uploads them to the cloud, and alerts users if something is wrong.

When that day arrives, we'll have fulfilled the promise of ASHRAE 110's dynamic challenge—and taken it further. We won't just test containment once a year. We'll live in a world where containment is always visible, always verifiable, and always trusted.

Chapter 8
Tri-Color in Specifications

Why Specifications Matter

Laboratory design is driven by specifications. Architects, engineers, and EHS officers rely on written requirements to ensure consistency, performance, and compliance. If it isn't written into the spec, it usually doesn't get done.

Tri-Color, if properly specified, can demonstrate containment and preserve the actual testing. It can mitigate risk and reduce liability.

When Tri-Color is written into project documents, it moves from "optional" to expected.

The Spec Effect:

If it's not in the spec, it doesn't happen.
If it is in the spec, it becomes the standard.

The Gap in Current Specs

Most specifications today reference face velocity and ASHRAE 110. While important, both have limitations:

- Face velocity is a poor predictor of containment
- ASHRAE 110 provides valuable data but rarely includes dynamic challenges

The result? Laboratories that appear compliant on paper may still allow containment failures in practice.

Tri-Color fills this gap. It provides a visual, dynamic test that complements the numbers. By specifying Tri-Color, owners and planners can demand proof that hoods work safely in the real world, not just in a test room.

How to Write It In

Here's example language that can be adapted for project specs:

Section: Laboratory Fume Hoods – Performance Testing

Fume hoods shall be tested in accordance with ASHRAE 110-2016 (As Manufactured, As Installed).

Hoods shall be evaluated using the Tri-Color Airflow Visualization Protocol (Fume Hood Certified, 2025) to assess containment under dynamic conditions.

A modified ASHRAE 110 report shall include:

- Face velocity profile
- Smoke/haze visualization
- Tri-Color laser visualization (blue, red, green)
- Video documentation of all tests

A hood shall not be accepted until it demonstrates safe containment performance under both ASHRAE 110 and Tri-Color visualization.

Key Elements to Specify:

- Tri-Color protocol (with citation)
- Modified ASHRAE 110 report format

Video capture requirement
Dynamic challenges included

The Benefits for Stakeholders

- **Architects & Engineers** - A defensible way to demonstrate their design delivers safe performance and performs as specified
- **EHS Officers** - Clear, visual proof that hoods protect users
- **Owners** - Reduced liability and a safer investment
- **Users** - Confidence that the hood they're working in truly contains

When written into specs, Tri-Color becomes more than a test—it becomes a contractual expectation.

Case Example

At several universities where Tri-Color was specified, the difference was immediate:

Hoods that "passed" ASHRAE 110 but leaked under Tri-Color were identified and corrected before the lab opened

Facilities gained a permanent video record tied to each hood's HIN label

Users reported greater trust because they could see proof of containment, not just read numbers in a report

Closing Thought

Tri-Color belongs in specifications because safety should never be optional. By requiring it in the same breath as ASHRAE 110, we ensure every hood is tested not only for compliance, but for real-world containment.

Once Tri-Color is in the spec, it isn't just a new tool. It's the new standard.

As you can see from this photo, Tri-Color isn't just for fume hoods, anywhere you want to visualize airflow, you can use Tri-Color.

Chapter 9
Voices of Tri-Color

As the inventor of Tri-Color it gives me a great sense of gratitude when people see value in what you have done.

Lab Manager awarded Tri-Color the 2024 Innovative Safety Product of the Year.

"I have been testing fume hoods with Tri-Color for over a year now and what I love about Tri-Color is it allows me to visualize how the fume hood is performing….. it cuts the certification time in half" - *Gabriel Dobre Critical Air Technologies*

"We recently purchased the Tri Color system and it was invaluable in identifying cross drafts negatively impacting containment and turbulent flow on a high-density fume hood lab. The supply air distribution system was reworked to minimize room air flow effect on the fume hoods. Based on our use of the Tri Color" - *Clyde Harvey Systems Commissioning Specialists Inc.*

"I was able to better understand the inner workings of our fume hoods and improve upon our existing testing procedures with Tri-Color. It has also been a great training tool for staff and students!" - *John Brown University*

Conclusion
From Compliance to Containment

The question is no longer whether Tri-Color works. It does. The question is: **will you use it?**

Standards like ASHRAE 110 gave us the foundation. Tri-Color fulfills that intent by making containment visible, dynamic, and undeniable. Numbers and reports have their place, but they cannot replace proof you can see with your own eyes.

Now it's your move:

- **Specify it** in your next project.
- **Require it** in your next hood acceptance test.
- **Train with it** so every user understands containment.
- **Demand it** so safety is proven, not assumed.

Don't wait for standards to catch up. Labs can't afford to. Tri-Color is here, it works, and it's accessible to every facility that chooses to adopt it.

Because when airflow is visible, safety is undeniable.

Appendix A

TRI-COLOR
FUME HOOD VISUALIZER

Download the Tri-Color Testing Protocol

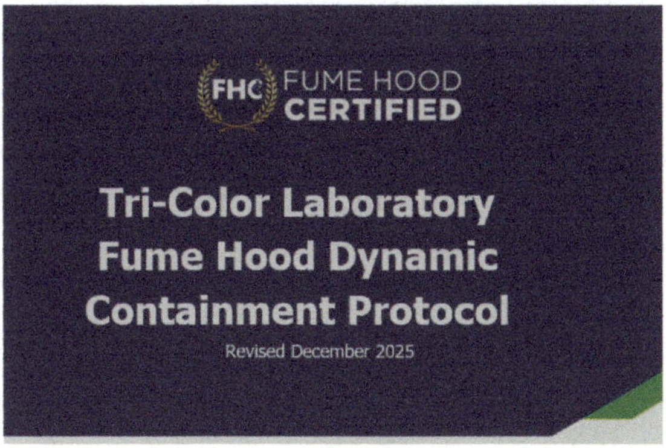

Published by Fume Hood Certified, LLC on December 1, 2025
This Protocol can be downloaded from our website
www.tricolorlasertesting.com/protocol

You can contact us at experts@fumehoodcertified.com our telephone is 623-562-0907 and our mailing address is PO Box 71477 Phoenix, AZ 85050 USA.

Special Note

This testing protocol is a voluntary procedure. Compliance is voluntary until and unless a legal jurisdiction makes compliance mandatory through legislation. Or it is specified in the scope of work in a laboratory project.

Our management team is available to answer questions and to address special conditions not addressed within. Contact us at experts@fumehoodcertified.com

Consensus standards take years to develop and update. Most ANSI standards are on a 10-year cycle. When Standards Lag, Innovation Leads. Standards take years to update. Innovation moves faster.

ASHRAE 110 has evolved, but slowly. Tri-Color testing already addresses challenges included in current standards. Waiting for standards means delaying safety improvements.

*Building on ASHRAE 110-2016, **Tri-Color Airflow Visualizer** adds dynamic elements as suggested in the standard. Additionally, Tri-Color can be used as an alternative to SF6 Containment Testing. A modified ASHRAE 110 report can be produced using the Tri-Color data. Our **Fume Hood Performance Tracker** online software provides a format for viewing actual test results.*

This protocol was developed by the experts at Fume Hood Certified. Since 2007, lasers have been effectively used to help visualize airflow. But like the standard ASHRAE 110, it took a highly trained technician to get any actionable results. In an effort to promote more fume hoods being tested in a meaningful way, the use of this laser enhancement had to be simple to perform and the results easy to interpret and understand. Tri-Color is simple and easy to use - and the results are very visual - so anyone can see how a fume hood is performing. Tri-Color has become the De Facto standard for dynamic containment testing.

DISCLAIMER
Fume Hood Certified uses its best efforts to promulgate Fume Hood Performance testing for the benefit of the public in light of available information and accepted industry practices. However, Fume Hood Certified does not guarantee, certify, or assure the safety or performance of any products, components, or systems tested, installed, or operated in accordance with this testing guideline or that any tests conducted under its guideline will be nonhazardous or free from risk.

(This foreword is not part of this standard. It is merely informative and does not contain requirements necessary for conformance)

FOREWORD

This voluntary testing protocol is built on enhancing ASHRAE 110 testing and as a replacement for simple average face velocity readings as a method for determining a fume hood's safe performance. The following is from the ASHRAE 110-2016 standard:

"It is important to evaluate the performance of the laboratory hood under dynamic conditions. This performance test method may be modified to evaluate a dynamic challenge. Specific operations, such as a pedestrian walking past the hood, laboratory doors opening, and specific actions at the hood, are only a few of the challenges that could be expected at the hood. This test method addresses only the dynamic challenge of sash movement. VAV hoods place a significant emphasis on the sash movement and the potential effect on hood performance. However, some constant-volume hoods may also experience a decrease in protection when the sash is moved."

Building on this, the Tri-Color Dynamic Challenges are a series of tests designed to visualize fume hood airflow and dynamically challenge containment.

Testing Protocols such as ASHRAE 110, EN 14175, and MD-15128 are excellent tests in many situations, they are also complicated, expensive, and require a well-trained tester. It is also difficult to really understand the reports generated from the test data.

Tri-Color has an added element; it allows you to visually see the results. If you can't see a problem, you can't fix it. Testing with Tri-Color is simple, affordable and you can see the results in real-time. Tri-Color has been developed to be a DIY Test or to be performed by a trained and certified Tri-Color Tester. The goal is to have more fume hoods tested on a regular basis. Testing is the only way to know how a fume hood is performing and whether it is providing the level of safety and protection it was designed to.

Building on the ASHRAE110 protocol for a face velocity profile and smoke visualization, along with the Canada MD-15128 standard, Tri-Color takes visualization to the next level.

CONTENTS

Testing Performance of Laboratory Fume Hoods Utilizing the Tri-Color Airflow Visualizer

Tri-Color offers an easy upgrade to the Fume Hood Certifications that rely on face velocity testing as the primary component. There is no direct relationship between face velocity and containment. In fact, the majority of fume hoods that fail containment testing have acceptable face velocities.

Tri-Color is also an alternative to the SF6 testing utilized in the ASHRAE 110 standard. Given that SF6 is a strong greenhouse gas and is expensive and difficult to obtain in many places, it should be reserved for troubleshooting.

Tri-Color makes fume hood testing safe and simple. In addition, it is an excellent educational tool because you can see the results - you will see cause and effect in real time.

To assist with recordkeeping and compliance, we have developed the HIN (Hood Identification Number) Program that gives each hood a unique serialization.

HOOD
IDENTIFICATION
NUMBER

23-R24 Y44

SCAN ME

FHC FUME HOOD CERTIFIED

The HIN number interfaces with our "**Fume Hood Performance Tracker**" software. Here, each hood has a home page and is supported with test data, photos, videos, and other related data and information. It is the intent of this protocol to develop a very visual collection of content to document the hood's performance over time.

TEST PROCEDURES

FACE VELOCITY PROFILE

Establish a uniform grid pattern across the face of the hood.

The cells shall be no more than 12" x 12".

Example:

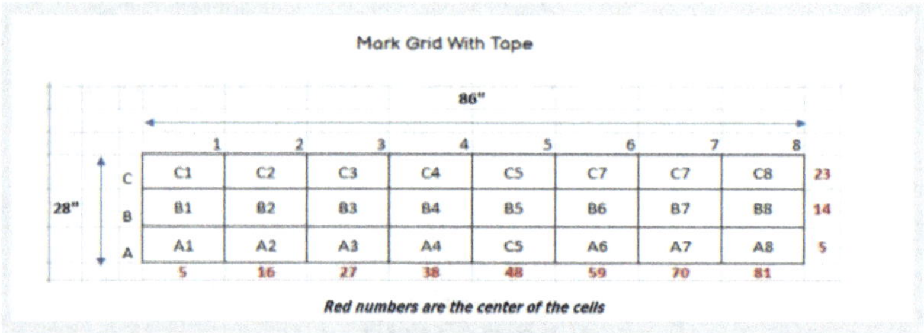

Purpose

The purpose of the face velocity profile is to understand turbulence. As air enters the sash opening it is entrained into the flow patterns within the fume chamber. Fume Hoods are very turbulent; it is this instability and dynamic airflow patterns that contribute to loss of containment.

There is no direct relationship between face velocity and containment. In fact, most hoods that fail containment testing (SF_6) have acceptable face velocities. Given that the classification of the fume hood is an "Exposure Control Device", we should focus on containment when evaluating fume hood performance - looking at just the average face velocity has no real meaning.

With the Face Velocity Profile, we can better understand what is happening at the face of the hood near the user's breathing zone.

In a perfect world, the readings from the various grid cells would be identical, but labs are not perfect worlds and room conditions greatly influence fume hood behavior and performance.

Procedure

Whether you are using a stand mounted hot wire anemometer or an array of hot wire transducers, they shall be mounted on a stand and not handheld. In each cell of the grid, take a minimum of 20 readings, one reading per second. Use this data to develop an average velocity for each cell.

This test is an indication of turbulence in the hood and a predictor of containment. The less variation from cell to cell the better the fume hood will contain.

Note: Capture photos of each test to show the setup. These should be uploaded to the Fume Hood Performance Tracker Software.

Full Open

Record reading from the center of each cell (20 readings @ 1 per second) and calculate the cell's average velocity. Once all cells have been recorded, calculate an overall average. Record the highest and the lowest cell's readings. All the cells should be within plus or minus 10% of average. The difference between the high and the low reading shall be less than 20%.

Sash Stop or 18" (Design Opening)

This test is the same as the full open test but is performed at the operating position.

All cells should be plus or minus 10% of average. The difference between the high and the low reading shall be less than 20%.

Note: Best Practices are for each hood tested to have a reference ASHRAE 110 AM test report as a baseline. The As Manufactured (AM) test was performed in a test lab with near perfect conditions. When field results vary significantly from the AM test results, it is an indication that there are issues with the room conditions.

Note: For the following exercises and challenges it is imperative to record ample video to adequately capture the performance results.

Please follow these dedicated video instructions:

1. **Placement of the video camera is critical to capturing loss of containment.**
    a. **Always mount the camera on a tripod or fixed mount - outside of the hood and off to the side.**

   b. **Place the video camera so that its line of sight is as parallel with the sash plane as possible.**
2. **The video camera will capture loss of containment much more effectively if the hood's interior light is turned off.**
3. **Maximum video quality will be achieved if the room's lights can be lowered and/or natural window light is minimized.**
4. **Record the video for each exercise/ challenge for a minimum of 30 seconds.**
   a. **When troubleshooting a hood, record for a minimum of 90 seconds. The green laser can be subtle, more recorded time is best.**
5. **All video should be uploaded to the Fume Hood Performance Tracker Software.**

SMOKE VISUALIZATION (AIRFLOW PATTERNS)

Purpose

This test is a visualization of a hood's ability to contain vapors. It consists of both a small local challenge and a large-volume challenge to the hood. The intent of this test is to determine the performance of the hood as it is typically used. Because the tester is often at the face of the hood while performing the tests, care shall be exercised to ensure that the body of the investigator does not influence the smoke visualization. Under ideal conditions, the smoke will flow smoothly, drawn from the point of release toward the slots in the rear baffle.

The following definitions describe typical airflow problems as demonstrated by smoke visualization:

- If the smoke remains on the work surface without smoothly flowing to the back baffle, the airflow is described as
  "lazy."
- If the smoke moves forward toward the front of the hood, the airflow is described as "reverse flow."
- A minimal amount of reverse flow will nearly always occur at the marine edge of the work surface and is considered normal.
- Reverse flow does not apply to the forward motion of the roll inside the hood that occurs in the upper cavity of the hood above the hood opening or to the cyclonic motion behind a closed horizontal sash.

2.1 SMALL VOLUME LOW VELOCITY SMOKE VISUALIZATION

Procedure

1. Place the sash in the test position.
2. Test the operation of the bottom airfoil by releasing smoke from the small smoke source under the airfoil.
3. For a successful test, the smoke shall be exhausted smoothly and not be entrained in the vortex at the top of the hood.
4. Discharge a stream of smoke from the small smoke source along both walls and the work surface of the hood in a line parallel to the hood face and 6 in. (150 mm) behind the face of the hood, and along

the top of the face opening. Carefully observe the action in the corners.

5. Release smoke along the work surface. Carefully observe the air flow behavior.
6. All air flow behavior shall be noted.
7. Release smoke above the bottom of the sash and inside the hood.
8. Observe the airflow behind the sash with specific attention to the airflow as the hood roll meets the bottom of the sash.
9. Release smoke outside the hood.
10. Observe the airflow into the hood and determine whether the room air currents appear to affect the airflow at the hood.
11. Where appropriate, observe the influence of air entering the hood between the sash and the header panel.
12. Release smoke in the cavity above the hood opening.
13. Observe the roll inside the hood.
    a. Pay particular attention to the clearance of the smoke, the slots the smoke enters, and any tendency of the roll to follow the inside of the sash toward the opening.

14. For "as used" (AU) tests:
    a. If hot plates or other heat sources are present, observe the influence of the heat sources on the airflow inside the hood.
    b. During the test, any lazy or reverse airflow observed may be caused by the equipment and should be noted and documented.
15. For horizontal sash or combination sash configurations:
    a. Open the horizontal windows and release smoke

along the inside of the sash near the open vertical edge of the sash.

16. Continue this exercise until you are satisfied that you have demonstrated this hood's behavior adequately.
17. Allow the haze to be exhausted by the hood when this exercise is completed.
18.

2.2 LARGE SMOKE VISUALIZATION

Procedure

1. A suitable source of smoke shall be used to release a large volume of smoke.
    a. Extra care in interpreting the observations is required because of the large volume of smoke and the momentum of the smoke. This is especially true when smoke is released outside the hood.
    b. Observation of the large-volume smoke release is often best done from the side of the hood face.
2. Release a large volume of smoke within the hood.
3. Pay particular attention to the air flow/smoke's behavior:
    a. Under the airfoil.
    b. Along the sidewalls.
    c. Along the work surface.
    d. Above the bottom of the sash.
    e. Release smoke in the cavity above the hood opening.
4. If equipment is in the hood, release smoke around the equipment.
    a. Observe the smoke's behavior

around the equipment.

5. For horizontal sash or combination sash hoods observe the smoke's behavior behind the sash.
6. During an "as used" (AU) test, equipment in the hood, such as heating devices and agitators, shall be operating to determine whether they contribute to leakage.
7. For ALL hoods:
   a. Release smoke outside the hood.
8. For ALL the locations and conditions above, the airflow/smoke's patterns shall be observed and noted.
9. All of the smoke generated within the hood shall be carried to the back of the hood and exhausted.
10. Continue this exercise until you are satisfied that you have demonstrated this hood's behavior adequately.
11. Allow the smoke to be exhausted by the hood when this exercise is completed.

TRI-COLOR AIRFLOW VISUALIZATION WITH CHALLENGES
(Loss of Containment)

Purpose

The purpose of Tri-Color is to have a fume hood containment test that is simple, economical, and the results are easy to understand.

Looking at a hood's performance on a scale of 1 to 10, with one being non-operational and 10 being a well performing hood against all the challenges, the closer we get to 10 on the scale the more robust the hood's performance is. The purpose of the challenges is to determine how robust the hood is.

Tri-Color Air Flow visualization is as much about education as it is testing. Given that you can see the relationship of what is happening in and around the hood and its impact on performance.

Procedure

Building on ASHRAE 110 smoke visualization, Tri-Color adds several new dimensions. This new approach begins with five fixed lasers - one blue, two red and two green. One additional green laser is included as a handheld laser to specifically identify air flow in any area in and around the fume hood.

BLUE - The blue laser is mounted directly onto the Tri-Color Airflow Visualizer inside the fume chamber. This blue laser is pointed vertically and highlights a cross-section of the fume chamber with emphasis on the vortex.

(Pay particular attention to the airflow behavior at the roof of the fume hood while being challenged)

RED - By using the magnets and corresponding mounting plates, affix the two red lasers on the inside of the hood's sash just above the sash handle - one laser on the left side and one laser on the right, the lasers shining across the sash opening to create a red light curtain just behind the sash plane.

Once adjusted, this will help visualize what is happening with the airflow just behind the sash and near the user's breathing zone. Once the haze is dispensed, the red lasers will dramatically highlight turbulent air flow.

Pay particular attention to the airflow that clings to the inside sash glass and near the sash handle)

GREEN - Mounted on the outside of the sash, similar to the red lasers, there are a pair of green lasers shining across the sash opening to create a. Once properly adjusted, these lasers create a green light curtain directly in front of the sash that highlights any haze escaping the hood (loss of containment). During the challenges, any loss of haze will generally be seen via the green color and indicate a potential problem. Of course, this loss of containment needs to be examined to determine how the loss has occurred.

(Pay particular attention to the micro-puffs of green that occur right near the lower edge of the sash handle and the vertical edges of the fume hood)

The sixth laser is a handheld green laser that allows you to focus on any area. (This can aid in identifying a very specific area of concern in or around the hood by directly aiming this green laser to that area)

3.0 FORCED FAILURE

Purpose

The purpose of the forced failure is to establish a baseline for what a failure looks like on the hood being tested. It confirms proper setup and allows you to capture a video showing a failure in containment.

Procedure

1. Place the Tri-Color Fume Simulator inside the hood and correctly connect the unit, allowing for a short warm up period.
2. Insert the 90-degree elbow into the fume simulator and point it directly towards the sash opening, this should cause a failure (loss of containment) and that loss will be highlighted by the green lasers. This will give you a baseline for a failure and ensures that the setup is correct.
3. Remove the 90-degree elbow when this challenge is completed.

3.1 FULL OPEN CONDITION

Purpose

The purpose of the full open condition is to determine the hood's performance level with the sash in the set-up position. A fully open sash is the worst case for fume containment. While the hood is not intended for operation in this position (setup only), it is important to understand how the hood performs in this sash position.

Procedure

1. Place the Tri-Color Fume Simulator inside the hood and correctly connect the unit, allowing for a short warm up period.
2. Ensure all the lasers are correctly in place and operational.
3. Dispense haze within the hood. Stand back from the hood so as not to disturb the airflow entering the hood.
4. Pay particular attention to the airflow around the sash opening, the rear baffles and the uppermost portions of the hood's interior.
5. Continue this challenge until you are satisfied that you have demonstrated this hood's behavior adequately.
6. Allow the haze to be exhausted by the hood when this challenge is completed.

Note: Room conditions are very dynamic and are ever changing. This sash position often shows a hood failing occasionally. This is why the ASHRAE 110 SF6 containment tests run 5 minutes. It is not uncommon for a hood to lose containment only a few times during that 5-minute test.

3.2 DESIGN OPENING

Purpose

This challenge demonstrated how the hood is performing during normal operation. The design opening is the working sash height that the system was designed for. The design opening is often approximately 18″; many hoods have sash stops at this design position. Some hoods have matching arrows at the design position. This should be the position the sash is opened to when working.

Procedure

1. Place the Tri-Color Fume Simulator inside the hood and correctly connect the unit, allowing for a short warm up period.
2. Ensure all the lasers are correctly in place and operational.
3. Dispense haze within the hood. Stand back from the hood so as not to disturb the airflow entering the hood.
4. Check carefully for any signs of haze being detected by the green lasers, which would indicate loss of containment.
5. Close attention should be paid to the red highlighted haze at the bottom of the sash handle. This is the most common area for leakage to occur and is the closest to the user's breathing zone.
6. Continue this challenge until you are satisfied that you have demonstrated this hood's behavior adequately.
7. Allow the haze to be exhausted by the hood when this challenge is completed.

3.3 STANDING IN FRONT OF THE HOOD

Purpose

The purpose of this challenge is to validate the hood's performance level while the operator is standing in front of the hood. While it is possible for a hood to have loss of containment even when an operator isn't present, it is a much more common occurrence when a person is working in front of the hood. A person in front of the hood creates turbulence that can draw chemicals out of the hood.

Note: This challenge can be conducted with the sash in the design opening and/or the full open position.

Procedure

1. Place the Tri-Color Fume Simulator inside the hood and correctly connect the unit, allowing for a short warm up period.
2. Ensure all the lasers are correctly in place and operational.
3. Open the sash to either the design opening or full open position - depending on which challenge is being administered.
4. Dispense haze within the hood.
5. Stand directly in front of the hood as if you are the operator. Stand still with your arms slightly extended into the hood as if to be reaching for an object.
6. Check carefully for any signs of haze being detected by the green lasers, which would indicate loss of containment.
7. Close attention should be paid to the red highlighted haze at the bottom of the sash handle. This the most common area for leakage to occur and is the closest to the user's breathing zone.

8. Continue this challenge until you are satisfied that you have demonstrated this hood's behavior adequately.
9. Allow the haze to be exhausted by the hood when this challenge is completed.

3.4 RAPID SASH MOVEMENT

Purpose

The purpose of this challenge is to determine the hood's performance level during rapid sash movement. As the sash(es) are opened and closed they cause movement in the vortex within the hood. On a vertical sash, the vortex goes up and down with the sash. When there is rapid sash movement, it is possible for the sash to move faster than the vortex causing the potential for loss of containment.

Procedure

1. Place the Tri-Color Fume Simulator inside the hood and correctly connect the unit, allowing for a short warm up period.
2. Ensure all the lasers are correctly in place and operational.
3. Lower the sash to the closed position.
4. Dispense the haze within the hood.
5. While standing directly in front of the hood, raise the sash at a faster-than- normal rate to the design opening.
6. Check carefully for any signs of haze being detected by the green lasers, which would indicate loss of containment while the sash is in motion - especially at the edges of the sash.
7. Now, lower the sash to the closed position at a faster-than-normal rate. Again, check carefully for any signs of haze being detected by the green lasers - especially at the edges of

the sash.

8. Continue this challenge until you are satisfied that you have demonstrated this hood's behavior adequately.
9. Allow the haze to be exhausted by the hood when this challenge is completed.

3.5 COMBO SASH/ HORIZONTAL SASH

Purpose

The purpose of this challenge is to measure how well the hood performs with the sashes in multiple positions. Combination or Horizontal sashes create unique airflow conditions. Horizontal sash openings can create vertical multiple vortices; these vortices create significant airflow disruptions that are more prone to loss of containment than those from a traditional vertical sash design, especially along the vertical edges of the sash glass.

Procedure

1. Place the Tri-Color Fume Simulator inside the hood and correctly connect the unit, allowing for a short warm up period.
2. Ensure all the lasers are correctly in place and operational.
3. Lower the sash to the closed position (horizontal sashes) or lower the sash to the closed position (combo sash).
4. Determining the hood's level of performance will require moving the various sashes into multiple positions in order to detect if the hood is having difficulty containing.
5. Horizontal Sash - adjust the sashes to the positions below and check carefully for any signs of haze being detected by the green lasers, which would indicate loss of

containment - especially at the edges of the sash.

   a. Two (2) Sliding Sashes
      i. Left Opening
      ii. Right Opening
   b. Three (3) Sliding Sashes
      i. Left Opening
      ii. Center Opening
      iii. Right Opening

6. Combo Sash - Lower the entire sash to the closed position and adjust the sashes to the positions below.
   a. Two (2) Sliding Sashes
      i. Right Opening
      ii. Left Opening
   b. Three (3) Sliding Sashes
      i. Left Opening
      ii. Center Opening
      iii. Right Opening
7. Check carefully for any signs of haze being detected by the green lasers, which would indicate loss of containment - especially at the edges of the sash.

NOTE: Vertical edges of horizontal and combination sashes tend to create vortices at those locations which can cause loss of containment. The sash edge vortices combined with the vortices created at the roof of the sash are especially challenging for these hoods.

8. Continue this challenge until you are satisfied that you have demonstrated this hood's behavior adequately.
9. Allow the haze to be exhausted by the hood when this challenge is completed.

3.6 WALK-BY

Purpose

The purpose of the walk-by challenge is to examine the effects of moving pedestrian traffic while the hood is in normal operation. This walk-by is a very difficult challenge; even the most robust hood can show loss of containment while passing all other challenges. Pedestrian traffic in front of the hood can easily draw chemicals out of the hood as the body passes by. The purpose of this challenge is to replicate pedestrian traffic and movement in front of the hood. People moving by create low pressure areas behind them that can draw fumes out.

Procedure

1. Place the Tri-Color Fume Simulator inside the hood and correctly connect the unit, allowing for a short warm up period.
2. Ensure all the lasers are correctly in place and operational.
3. Raise the sash to the design opening position.
4. Dispense haze within the hood.
5. Have a walking pedestrian pass parallel along the full face of the hood at a normal walking pace.
6. Check carefully for any signs of haze being detected by the green lasers, which would indicate loss of containment, while the pedestrian is in motion - especially at the edges of the sash.
7. Have the pedestrian repeat the pass from the opposite direction.
8. Check again for any signs of haze being detected.
9. Continue this challenge until you are satisfied that you have demonstrated this hood's behavior adequately.

10. Allow the haze to be exhausted by the hood when this challenge is completed.

3.7 RAPID HAND MOVEMENT

Purpose

The purpose of this challenge is to demonstrate and document the effects of rapid hand movement. Rapid hand movement inside the fume chamber can disrupt airflow, create turbulence and cause loss of containment. Our goal is to see if the hood is robust enough to maintain containment during rapid hand movement while reaching into the hood and removing an object.

Procedure

1. Place the Tri-Color Fume Simulator inside the hood and correctly connect the unit, allowing for a short warm up period.
2. Ensure all the lasers are correctly in place and operational.
3. Raise the sash to the design opening position.
4. Dispense haze within the hood.
5. While standing directly in front of the hood, reach into the hood as if to be making adjustments or moving objects within the hood in a normal and customary way.
6. Check carefully for any signs of haze being detected by the green lasers, which would indicate loss of containment, especially around the operator's hands and arms. Pay particular attention to loss of containment in and around the breathing zone of the operator along the lower edge of the sash handle.
7. Repeat the challenge at various positions along the face of the hood.
8. Continue this challenge until you are satisfied that you

have demonstrated this hood's behavior adequately.

9. Allow the haze to be exhausted by the hood when this challenge is completed.

3.8 MATERIAL EXITING HOOD

Purpose

The purpose of this challenge is to determine the hood's performance level when moving objects in and out of the hood. When removing items from the hood, it is possible to drag chemicals out of the hood and pull these escaping chemicals directly into the operator's breathing zone.

Procedure

1. Place the Tri-Color Fume Simulator inside the hood and correctly connect the unit, allowing for a short warm up period.
2. Ensure all the lasers are correctly in place and operational.
3. Raise the sash to the design opening position.
4. Dispense haze within the hood.
5. While standing directly in front of the hood, move objects (i.e. flasks, scales, stands or similar items or props) in and out of the hood in a normal and customary way.
6. Check carefully for any signs of haze being detected by the green lasers, which would indicate loss of containment, especially around the operator's hands, arms and the objects. Pay particular attention to loss of containment in and around the objects as they are moved through the sash plane.
7. Repeat the challenge at various positions along the face of the hood.

8. Continue this challenge until you are satisfied that you have demonstrated this hood's behavior adequately.
9. Allow the haze to be exhausted by the hood when this challenge is completed.

3.9 LAB DOOR TEST

Purpose

The purpose of the lab door test is to determine the hood's reaction when lab room doors are opened and closed. The lab is normally under slight negative pressure compared to the hallways. When the door is opened, air rushes in. This changes the room pressure and depending on how negative the hood pressure was, this change in pressure can cause loss of containment. Containment is greatly impacted by the air balance of the room.

Procedure

1. Place the Tri-Color Fume Simulator inside the hood and correctly connect the unit, allowing for a short warm up period.
2. Ensure all the lasers are correctly in place and operational.
3. Raise the sash to the design opening position.
4. Dispense haze within the hood.
5. While standing directly in front of the hood, raise the sash to the design opening position.
6. Have the lab's door opened and closed in a normal way.
7. Walk in and out of the opening, closing the door between exit and entrance.
8. Carefully check for any signs of haze being detected by the green lasers, which would indicate loss of containment while the lab door is being opened and closed.

9. Continue this challenge until you are satisfied that you have demonstrated this hood's behavior adequately.
10. Allow the haze to be exhausted by the hood when this challenge is completed.

4.0 OTHER HOODS OPENING AND CLOSING THEIR SASHES

Purpose

The purpose of this challenge is to determine the fume hood's performance when other hoods in the lab are opening or closing their sashes. Most labs have VAV (variable air volume) supply systems. Many hoods also have VAV controls. As other hoods in the system open and close their sashes, the pressure is always changing. This change in pressure can cause loss of containment when the room's balance shifts.

Procedure

1. Place the Tri-Color Fume Simulator inside the hood and correctly connect the unit, allowing for a short warm up period.
2. Ensure all the lasers are correctly in place and operational.
3. Raise the sash to the design opening position.
4. Dispense haze within the hood.
5. Randomly open and close other lab hoods, at normal speeds, leaving the sashes open long enough for the test hood to react fully to the change in pressure.
6. Carefully check for any signs of haze being detected by the green lasers, which would indicate loss of containment while the other sashes are being opened and closed.

7. Continue this challenge until you are satisfied that you have demonstrated this hood's behavior adequately.
8. Allow the haze to be exhausted by the hood when this challenge is completed.

AS MANUFACTURED (AM), AS INSTALLED (AI) AND AS USED (AU)

As Manufactured (AM)

AM identifies an "as manufactured" test, that is, the laboratory hood is built and assembled by the manufacturer and testing is performed at the factory. It is very important to understand that AM testing is accomplished under "perfect" conditions within a manufacturer's hood testing room. The purpose of this test is to evaluate the hood's design characteristics solely. There are no outside influences such as diffusers, cross drafts, temperature issues, pedestrians, etc. that can influence the hood's performance. The results of this test may have no correlation to the hood's performance once installed within an actual laboratory ventilation system.

As Installed (AI)

AI identifies an "as installed" test, that is, the laboratory hood is installed at the location of the customer. The hood is tested empty, but with the ventilation system in the installation balanced and the hood in its final location.

As per the ASHRAE 110 definition, this is an empty hood and is intended to verify the integration of the hood with the overall laboratory ventilation system. Since the fume hood is not a standalone device, its performance is highly dependent on room conditions. Using the AM test as a baseline, the AI results should mimic the AM results. If the AI results differ greatly from the AM results, this is an indication of a problem with the room conditions, not the hood.

As Used (AU)

AU identifies an "As Used" test, that is, the test is conducted after the hood has been installed and used by the chemist. The typical equipment remains in the hood and other activities in the laboratory continue.

In the **As Used (AU)** test, the concept is to mimic how the hood is used during its normal operation. The equipment inside the hood remains and the test is conducted accordingly.

Operator behavior is also a major component of the hood's ability to contain. Many of the challenges described earlier are intended to duplicate an operator's normal behavior to determine the hood's performance during the AU test. About 25% of fume hood failures are caused by work practices.

Using the AI test as a baseline, the AU results should mimic the AI results. If the AU results differ greatly from the AI results, this is an indication of a problem rooted in the equipment load of the hood, the operator's work habits and/or related activity, not a hood problem.

Note: The Tri-Color Visualizer has an available AU adapter that allows the main Tri-Color unit to remain outside the hood and not impede any of the equipment that is inside the hood.

TEST MODIFICATIONS TO BETTER ACCOMMODATE SPECIAL SITUATIONS

There is a huge diversity of fume hoods with many different sizes and sash configurations. There are also many special applications fume hoods. Regardless of the situation, the hood is designed to minimize harmful user exposure to the chemicals in the fume chamber.

This protocol is written for a benchtop fume hood. Changing the configuration of the equipment or the execution of the challenges is acceptable for conducting modified Tri-Color tests.

APPENDIX A: RATIONALE FOR OBSERVATIONAL ASSESSMENT OF CONTAINMENT

A.1 Purpose

This appendix provides justification for the use of qualitative, visual observation methods (Tri-Color Visualization) in the evaluation of laboratory fume hood performance. It addresses common concerns regarding the qualitative nature of Tri-Color testing and establishes why direct observation of containment loss may, in many circumstances, be more actionable than quantitative data reported from standardized tracer gas tests (e.g., ASHRAE 110).

A.2 Background

ASHRAE 110 and other quantitative tracer gas protocols provide standardized, reproducible methods to measure containment under controlled test conditions. The results are reported numerically (e.g., average face velocity, tracer gas concentration at the mannequin's breathing zone). While these values are useful for benchmarking and compliance, they do not always translate directly to actionable information for end users.

Tri-Color Visualization, by contrast, is an observational method designed to demonstrate containment performance dynamically and intuitively. When loss of containment is observed, it is immediately evident to all observers—regardless of technical expertise—and can be addressed without further interpretation.

A.3 Justification for Observational Methods

A.3.1 Immediacy and Clarity

- Visual evidence of containment loss is direct and unambiguous.
- All observers, including laboratory personnel without technical training, can readily identify when vapors escape into the user's breathing zone.
- No interpretation of numerical data or reference to threshold values is required.

A.3.2 Actionability

- Observed containment loss prompts immediate corrective action (e.g., lowering sash height, reducing cross-drafts, adjusting equipment placement).
- Numerical results, while precise, often require specialized knowledge to interpret and may not lead to timely corrective measures.

A.3.3 Real-World Relevance

- Containment is influenced by dynamic conditions (e.g., door openings, personnel movement, equipment-generated turbulence).
- Standardized quantitative tests are conducted under controlled, repeatable conditions and may not reflect actual use scenarios.
- Tri-Color Visualization exposes these dynamic effects, providing a more accurate representation of real-world performance.

A.3.4 Prevention of False Assurance

- A fume hood may meet quantitative acceptance criteria but still fail to contain vapors under typical laboratory use.
- Observed containment loss removes ambiguity: if containment is visibly breached, the condition is unsafe, regardless of numerical test outcomes.

A.4 Conclusion

Tri-Color Visualization should not be considered inferior due to its qualitative nature. On the contrary, direct visual confirmation of containment loss provides a more immediate, intuitive, and actionable assessment of risk. When used in conjunction with quantitative protocols such as ASHRAE 110, Tri-Color enhances the overall safety evaluation framework by bridging the gap between standardized test results and real-world performance.

Appendix B

Download the ASHRAE 110 Tri-Color Sample Report

Free Fume Hood Stickers for Safety Training

If you would like some of these labels reach out to experts@fumehoodcertified.com

Get Hoodie Gear

https://fumehoodtesterscatalog.com/collections/hoodie-gear

Take advantage of Fume Hood Certifier's extensive resources!

fumehood.academy
Information regarding our in-person workshops in Phoenix. A great event for any experience level.

fumehoodcertified.com
Learn about all that FHC has to offer and how we may able to help you

tricolorasertesting.com
Home of the Tri-Color Visualizer system. Learn how we utilize a modified ASHRAE110 large volume test to challenge fume hood's dynamic containment abilities.

fumehoodonlinecourse.com
Chip's last book, "Laboratory Fume Hoods Explained" converted into a free online course

fumehoodperformance.com
The home for all your fume hood testing data for easy sharing, review and analysis. Works seamlessly with our Q200 and Face Velocity Profiler data capture software

fumehoodsafety.com
We host several introductory level videos here for students and employees that are new to fume hoods to get a understanding of safe user practices

fumehoodsoftware.com
Learn all about our Fume Hood Performance Tracker here, includes a walk-through video and a live demo.

fumehoodtesterscatalog.com
A one stop shop for everything you need to perform ASHRAE110 compliant fume hood testing, from smoke testing to Q200 factory calibrations, we have it all!

fumehoodtestingnetwork.com
Directory of industry professionals that utilize our Tri-Color testing system.

Safety is No Accident (Blog)
https://fumehoodtesterscatalog.com/blogs/lab-safety-digest

YouTube
www.youtube.com/@FumeHoodCertified

LinkedIn
www.linkedin.com/company/fume-hood-certified-llc

Facebook
www.facebook.com/fumehoodcertified

Laboratory Fume Hoods Explained (Free Online Course)
www.fumehoodonlinecourse.com

Bulk Sales

Help Spread the word.

We are offering large discounts for multiple copies. Reach out to us at experts@fumehoodcertified.com

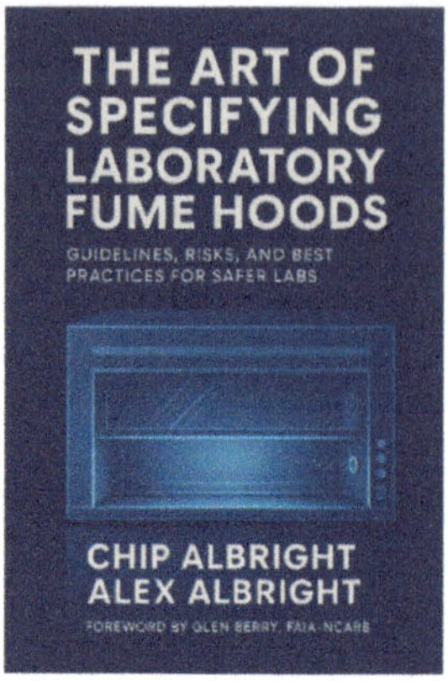

Do you have a Medical Portal?

Just like people use medical portals to monitor their health, laboratory fume hoods now have their own health tracker- **Fume Hood Performance Tracker**.

It's your digital portal to ensure every fume hood is operating safely and efficiently.

Fume Hood Certified is offering a **free trial** to help you get started. Discover how easy it is to track performance, flag issues before they become problems, and document safety like never before.

Don't wait for a safety incident to start monitoring your hoods-be proactive.

Email us today at **experts@fumehoodcertified.com** and start your free trial of the Fume Hood Performance Tracker.

HOOD
IDENTIFICATION
NUMBER
23-R24 Y44

SCAN ME

FHC FUME HOOD CERTIFIED

Your lab's safety deserves the same attention as your own health. Let us help you make that possible.

Made in United States
Orlando, FL
04 November 2026

90202241R00049